Raptor!

Life and Death of a Legendary Hawk in the Pacific Northwest

by Andy Norris

Illustrated and formatted by Eva Cantos

Copyright 2017 by Andy Norris
All rights reserved. No part of this book
may be transmitted in any form by any means
without permission in writing from the publisher.

Cover and illustrations designed by Eva Cantos

First printing January 2017

ISBN: 978-1490920924

For Yukon

4

350 Years Ago in the Pacific Northwest...

Chapter 1

The two ravens had been watching the nest for an hour. High atop a five hundred year-old cedar, hidden behind converging boughs of flattened evergreen leaves, the two juvenile males stood motionless on a moss-covered branch peering at the nest through a small break in the foliage. Although they had come to rest near the nest quite by accident, they were now preoccupied with it and took care not to reveal themselves.

The ravens had set out from the Puget Sound several days ago, having spent the first year of their lives on one of the larger islands. Their departure may have originated after a final, harsh scolding from a parent, from a lack of food, or perhaps from a simple desire to quench a yearning millions of years in the making – the internal push to establish a home range all their own. The two brothers left with no plan but to head south, and were now two hundred miles from their birthplace, having flown the distance in two days. Following a shoreline route, they had craftily navigated headwinds while scavenging for food on the beaches. They explored the perimeters of several estuaries and had followed many river valleys to their end in their search for available territory. Thus far, they had come up empty: All territories suitable for life as a raven had existing claims, with many areas having been under familial control for generations untold. It was while scouting one of these occupied river valleys in the foothills of Oregon's coast range that the two ravens came to rest on the old cedar, seeking only temporary respite for their tired wings.

 They heard the shrieks and screams of the chicks before they saw them. With a week to go before fledging, the young strikers' voices had reached a competency their wings had not. When approaching the nest with food, the parents call out cak! cak! cak! cak! cak! The chicks respond by battling for the distinction of having the loudest and most active response. This battle for nest supremacy ensured the strongest of the young would receive the most attention, a clear survival advantage for the most-fit individual and thus the species as a whole. However, like many evolutionary advantages, the associated disadvantage was lurking just beneath the surface, as the chick receiving the most attention was soon to find out.

The two ravens watched as the mother striker shrieked and landed on the branch of a nearby tree, where she shrieked some more. For several minutes she sat on the perch – a dead robin dangling from her talons - and watched the nest. Her chicks were going ballistic, and only after being satisfied with their vitality did she fly to the nest, where she was greeted with even more boisterous behavior. The loudest and strongest of the chicks grabbed the robin from his mother's talons and quickly started tearing off the downy abdominal feathers, which began gently falling toward the ground. Before the first feather had reached the forest floor, the hungry chick had consumed the meat of one breast and was working on the first of the two eyes.

Despite being one and a half times larger, a raven is not equipped to kill a healthy adult striker. In turn, it is extremely rare for a striker to take a raven, though they often prey on the less robust crow. That said, the largest predominantly bird-eating hawk of North America is the same size as a raven and it regularly takes them as prey, as the two juveniles learned firsthand with the loss of their sister to a goshawk several months ago. No, tangling with the mother hawk was not what the ravens had in mind. Neither did they seek to quell an acute hunger, though hungry they were. Something, however, was telling them they had stumbled upon a rare opportunity and the young of this hawk must be killed. Though the two brothers may not have been aware of it, witnessing the death of their sister had a profound effect upon them.

    Satisfied that the dead robin was on course to be consumed by her nestlings the mother striker set out to forage once again. In the effort to feed her rapidly growing chicks she he had not only been working her established routes several times a day but had been exploring new areas as well. She knew her hunting routes with remarkable accuracy and could safely navigate cliff edges, tree trunks and huckleberry bushes at thirty miles-per-hour relying only on her memory to keep from smashing into potential obstacles in the airspace ahead. Unfamiliar hunting areas carried a heightened risk of misfortune, but provided additional foods in times of great need. And so, with the angry look permanently affixed to the faces of all birds of prey, off she went to work unfamiliar treetops in search of finches and warblers.

    The raven brothers knew they had to act quickly. One at a time they silently dropped from the branch with wings outstretched and glided toward the nest. The robin carcass had been picked of its most nutritious parts and the brother and sister of the biggest chick were salvaging all they could from the leftovers. Seeing movement out of the corner of his eye and thinking it his mother with more food, the big chick shrieked and began flapping his wings. When a raven landed on mother's spot at the edge of the nest the young hawk froze. In two hops, the raven closed the distance between them and jabbed its pointed bill into the big chick's eye, piercing it. The big chick struggled toward to the other side of the nest, climbing over his brother and sister on the way. It was while he was on his sister's back that the second raven struck, poking and prodding him with his bill until he fell over the edge of the nest to the ground below. The raven then looked at the female chick, aimed his pointed bill at her eye, and was about to strike, when the screams of the mother striker caused him to look up.

Upon leaving her nest, the mother had stopped in a nearby tree to listen for the direction a flock of finches was heading. When she heard the shriek of the big chick, she immediately flew back and now came slicing through the air, furious, and before the raven could finish three strokes of his wings in retreat, her talons had pierced his abdomen. Tangled, the raven and hawk tumbled toward the ground. The raven was vigorously flapping in order to gain control of the fall and his wing was fully outstretched when it hit the alder branch. The wing snapped upon impact and would never work again. Before hitting the ground, the mother hawk released. Quickly righting herself and satisfied with the raven's debilitated condition, she began looking for the second raven, screaming all the while.

But the second raven was long gone, having flown away as fast as his wings would allow. He flew until just before sundown, at which time he descended to investigate a deer carcass that held promise of a much-needed meal. In the days to come, he would roam in an easterly direction, traverse the Cascade Range by early summer, and eventually settle in a remote desert canyon along a salmon-bearing river. He would feed on small mammals, insects and carrion for most of the year, but when the salmon ran in the spring, their eggs and carcasses would be the staple. He would fail to breed for six years, eventually taking a single mate for his remaining sixteen years and rearing twenty one young to adulthood before his death of natural causes.

From the nest high in the tree the mother hawk looked at the ground below and could see the damaged eye of her biggest chick. Though he would scream and shriek for days to come she would never again bring him food, and he would eventually die where he sat. The following day and only one hundred fifty feet away, nestled between clusters of sword ferns, the raven would take his last breath, dying from a combination of infection, dehydration and starvation.

With the absence of the big chick from the nest, the brother and sister began growing at a faster rate. The female was able to out-compete her brother for food and she became the dominant chick. This ability to dominate, along with her strength, outstanding eyesight and agility, were the characteristics most responsible for her future success as a bird-hunter. Because this young hawk embodied all of these characteristics in supreme form, we shall give her the name Raptor.

14

Chapter 2

Raptor was asleep when her father returned with a song sparrow dangling from his talons. Eight days had passed since the raven brothers' raid and in that time Raptor and her brother had experienced a final growth spurt and were now ready to leave the nest. However, they would not leave the comfort and safety of the only world they had known without some coaxing, so their father stopped on a nearby branch, the lifeless sparrow pinned beneath him. The parents would no longer deliver easy meals to their young but require they give chase for the prize. In doing so the young would not only leave the nest, but also learn to fly and eventually be able to catch food on their own.

The screaming of Raptor and her brother was almost too much for their father to bear, but he sat patiently on the branch, waiting. Raptor moved to the edge of the nest and began flapping her wings. She wanted to fly to her father and take the sparrow, but had only watched other avian creatures succeed in the act of flying. Propelled by hunger and instinct, she finally leapt off the edge of the nest and began to fall. She flapped clumsily at first, quickly sensed the direction her wings propelled the bulk of her body, began coordinating the action, and before she had fallen ten feet was in enough control to remain airborne. Moments later she was standing on the same branch as her father. After witnessing his sister's success the male fledgling moved to the edge of the nest and jumped. Learning to fly was not as easy for the brother and he landed awkwardly on the top of a huckleberry bush, shrieking all the while. After finding a strong branch from which to launch, the male jumped and flapped his wings, became airborne, and was soon standing a few feet from his sister.

Instead of giving up the sparrow, the father flew into the woods calling out loudly. Raptor and her brother followed, shrieking as they had in the nest. With each stroke of their wings flight became sturdier and in a short amount of time the fledglings' speed and agility almost matched that of their father. Only after being satisfied with their effort did he stop at the top a lightening-hit Sitka spruce.

Seconds after the fledglings landed on the spruce the father let go of the dead sparrow and it dropped towards the ground at free-fall speed. Without hesitation the young gave chase. It was Raptor who caught the sparrow in her talons, having come upon it twenty feet from the forest floor. She took the prize to a sturdy branch and began removing feathers and tearing into flesh. The brother, who had come up empty, flew to his father and cried out *cak! cak! cak! cak! cak!* The father immediately took off to work a hunting route and the young male followed. This game of chase and feed would go on for the next few weeks, with the chicks increasingly able to keep up with their parents. This brief period of learning from their parents, along with instinct and physical ability is all the fledglings would need to become great hunters of the sky.

For several weeks the young strikers remained in their parents' territory, feeding primarily on nestlings and fledglings. Small birds would remain the dietary staple throughout the hawks' life, though they would soon become practiced enough to take an occasional red squirrel from a tree or mouse off the ground. Raptor liked to work the edges of fields and became very good at hunting robins and meadowlarks, young and old alike. Her brother rarely left the woods those first few weeks, feeding on thrushes, sparrows and juncos, and once a lone band-tailed pigeon preoccupied with gorging himself on red elderberries. In time, brother and sister would both become expert hunters of many types of birds, but there was much learning to be done first.

Months prior, as the earth began its summer tilt towards the sun and the temperate rainforest had started to warm, the vernal rains arrived. By the summer solstice, the rains had yet to yield. For six days drenching rains had been falling without respite, and the creeks draining the coastal mountains had doubled in width and depth and were running fast, carving their routes deeper into loamy hillsides and subterranean chambers. For the striker family, rainy days offered a mixed bag of advantages and disadvantages. More calories were required on cold, wet days than on warm, sunny days, and that meant – all else being equal - more prey had to be caught during rainy periods. The hunting dynamic also changed with the wet spells and the hunters had to make adjustments. The sound of the water falling on leaves and the forest floor meant the warning calls of prey would not travel as far – an advantage – but that in turn meant the family had a more limited auditory range within which to hear prey. The motion of leaves and branches caused by the wind and falling raindrops meant the family's movements would be more difficult to decipher than in a static landscape

– an advantage – but that in turn meant they would have more difficulty picking out the movements of their prey. Most importantly, while the family had to make adjustments for acquiring food during the rains, so did their prey, and the family had to learn those behavioral changes as well. This rainy spell had been average in terms of prey acquired and the family had consumed one red squirrel, two deer mice, and nineteen perching birds in that time.

Three weeks had passed since the young strikers had fledged and they were now on the cusp of adulthood. There was no evolutionary advantage for the striker family to remain together and this would be their final day in one another's company. This particular morning saw a break in the clouds and the forest alive with the calls of countless birds taking advantage of the warming sun. The father striker was the first to stretch his legs and take leave. He dropped out of the tree and flew off, heading toward the estuary in a southerly direction. Though he left without fanfare or disagreement, his family would never see him again. His fatherly duties had been successfully accomplished and he simply left. He traveled south to northern California and spent the fall and winter working a fruitful riparian corridor along a tributary of the Eel River. He found a roost high atop an old redwood tree and remained in the area until late February, when he began wandering again, this time in a northerly direction. In early

March he passed over the estuary at fifteen hundred feet but did not stop. Several weeks later he began establishing a territory near a Haida Village on a large island nestled between the Gulf of Alaska and the Olympic Peninsula.

Despite the rain break and resultant activity of the birds of the forest, the three remaining family members had a tough day trying to acquire food. The mother's main hunting route covered a variety of territory - treetops, thickets, fields, shore line - and was usually very productive. For reasons quite routine – a successful evasive maneuver here, a timely warning call there – on this day, all targeted prey had managed to evade the hunters. Making things more difficult on the family was the fact that most nestlings had fledged. No longer were easy meals scattered in nests throughout the forest, but all prey had to be aggressively pursued. Having thus far come up empty, the family headed down a riparian corridor to the swamp at its outlet, hoping to come upon newly fledged redwing blackbirds.

Weighing forty-eight pounds and measuring almost four feet long, the Chinook salmon had been at sea for seven years. A veteran of nearly two dozen trips traversing a large arch across the north Pacific from Monterey Bay to the Gulf of Alaska to Kamchatka and back down again, the old fish had just left the ocean for good and was now lumbering up the creek the striker family was heading down.

The Chinook's length was at times greater than the creek's width, but he swam on. Having started his life in the creek as one of three thousand eggs laid by his mother and fertilized by his father, he was the only member of his familial pod to survive to breeding age. Most of his siblings had been consumed by predators while they were only a few inches long and still in the river. Herons, kingfishers, otters, raccoons, mink and larger fish of many species were among those who ate them. The survivors were predated upon over the following years by larger predators in the ocean, with sea lions, seals, dolphins, sharks and orcas being the primary beneficiaries of the bounty. The healed-over bite marks of a harbor seal paid testament to a close call the old Chinook once had, but thanks to his overall health and strength, outstanding ability to enact evasive maneuvers, and the tendency to position himself in the safer inner regions of the pod, he had survived and his genes would be passed on.

    The intense pressure to return to the place of his birth was millions of years in the making. The old Chinook found his way to the creek using olfactory cues and other homing mechanisms, and had timed his ascent upriver with the drenching rains so as to take advantage of the easier travel offered by the heightened water level. Despite the timing of his ascent, he would receive many scrapes on his body and twice suffer blunt trauma while attempting to navigate shallows and leap waterfalls up to ten feet high. Once reaching the exact pool where his life began, ninety-seven river miles from the ocean and sixteen hundred feet high in a saddle of the coast range, he will fertilize the eggs of as many females as he is able before dying of old age within one hundred feet of the exact place of his birth. His body will take months to decompose in the cold, temperate rainforest, as it becomes food for many kinds of life, including his offspring.

With ninety-plus miles of river travel ahead of him, the old salmon thrashed his tail back and forth and thrust off toward home. His sudden burst of energy startled a submerged water oozle and with a few flaps of her wings while underwater she fled from his path. The oozal – the only perching bird to forage for food while completely underwater – was the size of a robin and had reason to fear a fifty-pound fish that would feed on just about any living thing that would fit in his mouth. The oozal, too, would feed on whatever living things would fit in her mouth, which at this moment were salmon eggs. A pod was spawning a half-mile upriver and many eggs had failed to remain in the nursery, instead floating downriver and becoming deposited in a back eddy. The eddy was in the oozal's territory and she had been feeding on the eggs since sun-up. After being spooked by the large salmon, the oozal popped up from the bottom of the stream and flew to a rock beneath the waterfall where she had reared this year's brood, and began bobbing her body up and down and chattering. While the oozal was underwater gathering the salmon eggs, mother striker, with kids in tow, was speeding down the creek towards the swamp. Just as the spooked oozal emerged from the water, mother striker saw her, banked to the left, honed in on the bird, and tucked into a dive. At the last second the oozal saw the predator, executed an evasive maneuver and plopped back into the water. Mother was too slow to recover and slammed into the rock at an unusually fast speed. The wind was knocked from her lungs and her chest received a fracture. More significantly she had overextended the tendon in her heel, a severe blow to a bird that relies on a strong grip to hold and asphyxiate prey. For several minutes mother striker sat on the rock catching her breath and standing on one leg. Eventually mother took to the air and came to rest on a low branch of an alder. She stretched her leg but couldn't get it quite right. The fledglings stayed nearby, watching, trying to understand why their mother aborted the hunt.

    That evening, Raptor and her brother left the estuary and would never again return. Their leaving may have had something to do with their mother's debilitating accident or the day's unsuccessful hunt. Perhaps it had something to do with the departure of their father that morning, or it may have simply been a response to an instinctual cue millions of years in the making. Whatever the reason, the two young strikers chose to go elsewhere to live out their lives as great hunters of the sky.

Chapter 3

After leaving her mother's territory, Raptor flew over the headlands to the south and into the neighboring estuary. For several weeks she worked the area, had great success acquiring food, and did not encounter any hawks of her kind. The resident red-tailed hawk harassed her on two occasions, but she simply disappeared into the undergrowth, the red-tail being too cumbersome to follow. The highland forests, cascading riparian corridors and swamplands adjacent to this estuary were thus perfectly suited for a striker and would become her territory for years to come.

Raptor established hunting routes throughout her range, becoming more familiar with the habits of her prey with each tour. She learned that song sparrows could be found singing from various perches in the morning hours, then in the undergrowth for the remainder of the day. From sunup to sundown flute thrushes hunted insects from a low perch and often broke up the hunt with visits to patches of ripened salmonberry and thimbleberry. Flocks of cedar waxwings descended upon insect hatches and berry flushes alike, and Raptor encountered several flocks throughout her range with regularity. Stellar jays frequented the thick forests, while finches and warblers were often feeding in the treetops. At any time of day, redwing blackbirds could be found in the swamps. Raptor quickly noticed that robins went to sleep last, and often took one from a high perch while it was singing the day's final song, a half-hour after the sun dropped beneath the horizon.

While Raptor was learning to exploit the resources of her area, nine four week-old elk calves had weaned and were finding out which grasses, shrubs and trees were favored for consumption. Browsers, elk have a wide-ranging vegetarian diet, at times eating the leaves of trees and shrubs and other times grasses alone. Their thick tongues allow them to eat even the thorniest of plants, such as thimbleberry and salmonberry, as the smallest of the calves was just finding out.

The herd was presently following their established trails to the marsh, grazing as they walked, as was accustomed. The marsh grasses were in full bloom, giving the floor of the estuary a bright green color and it was toward them the herd was headed. Slowly, first one, then two, then in ones and twos, the entire herd of forty-nine emerged from the woods into the marsh, heads down, foraging all the while. The patriarch of the herd, a twelve hundred pound bull with antlers five-feet in length, stood apart from the rest. Raptor was transfixed and looking in his direction, though it wasn't the bull she was concerned with. Feeding on the ground a dozen yards beyond the bull was the small plover named killdeer. Raptor quickly dropped from her perch and flew fast and low. Reaching a speed of twenty-five miles per hour it looked as though Raptor was going to slam into the side of the bull. At the last second she rose up and over the bull, then swooped down towards the bird. The killdeer began screaming *killdeer! killdeer! killdeer!* and was only two feet off the ground before Raptor collided with him. Raptor easily wrangled the killdeer to the ground and spent the next several minutes asphyxiating him. The bull and a few nearby cows nonchalantly watched the killing while continuing to graze. When the struggle was over, Raptor took her prize to a small spruce just beyond the tide line and ate.

Hours later as she sat digesting the killdeer, Raptor witnessed one of nature's greatest hunters in action. Though presently full and unable to catch them besides – she had tried twice and failed miserably – Raptor had become preoccupied with the boisterous feeding of a family of green-winged teal. The speed and strength of a healthy duck's flight surpassed her own and she could not take one from the air. Likewise, taking a duck from the water required skills she did not possess, as only the eagle had perfected the art. None the less, the duck family's activity was the only motion in an otherwise static landscape, so Raptor watched.

Raptor had seen eagles hunt ducks several times. The hunt often began from a camouflaged perch high in an evergreen, several hundred feet up a slope on the south shore. Other times the hunt originated as a spontaneous event, the eagle coming upon a flock of ducks during a routine patrol of the territory. Once the ducks were spotted, the eagle would pick the appropriate approach. It might be a wing-tucked dive from high above, a stealthy drop over nearby treetops, or a line of travel that gradually moved him into striking range. Whatever the approach, the initial attack often came up empty, but that was of no concern to the eagle. The eagle's best chance lay in the ability to be patient, to soar, to hover, and most importantly, in his knowledge of duck behavior. Some ducks flew as the eagle approached while others dove into the water's depths. Those who flew would be safe; it is extremely rare for an eagle to pluck a duck out of the air. The ducks who chose to evade the eagle by ducking into the water (hence the name) might not be so lucky. Eagle, well practiced at catching fish, applied the same technique to the submerged ducks. Most ducks are able to stay underwater for a minute or more, and many use the opportunity to reach shore, hiding among piles of storm-blown driftwood or shrubbery. But those who remain water-bound aren't so fortunate. Eagles know a submerged duck has to breath eventually and so remain in the air, patiently hovering, soaring and watching. When the duck finally comes toward the surface to fill its' tired lungs with air, Eagle drops, talons extended, and plucks it from the water like a fish. But eagle was not the only avian duck hunter, nor was an eagle presently in sight.

  Raptor saw, a full half-mile away and perched six-feet up on a snag of driftwood the world's most ferocious and capable avian hunter. There this magnificent predator sat, watching the teal as well. Raptor had seen this tough bird before, harassing one of the six adult eagles that also call the estuary home. Like the elk, the tender shoots of marsh grasses had attracted the teal to the estuary and they were taking advantage of the bounty. The tide was going from low to high, and as soon as the grasses were out of reach, the teal would move to a more shallow area further toward the mouth of the bay. The peregrine falcon knew the rising tide would initiate their move and had chosen her perch accordingly.

Once the tender shoots were beyond the range of their outstretched necks, the teal began chattering and stretching their wings, just as the falcon had anticipated. The peregrine noticed this and began stretching her wings as well. After the teal's chattering reached it's crescendo, the family lifted off and headed down a drainage to the next feeding ground, keeping their flight path direct and just a few feet above the water. Having watched the teal family during the past two days, the peregrine knew not only where they were heading, but also the route they would take. She left her perch and in seconds matched the speed of the teal, took a low track, and was able to stay out of the teal's sight by flying just above the water yet below the tops of the marsh grasses. Despite being ever mindful of the threat predators posed, the teal could not see the falcon, and would not until it was too late to be of consequence.

This particular falcon was well practiced at taking ducks in the marsh and preferred pursuing them from a standstill. She had used the snag as a perch to begin several successful hunts, and the present effort would continue her outstanding record of success. The drainage the teal were following paralleled the drainage the falcon was speeding down, but was separated by mudflats a half mile wide. A quarter mile from where the birds' flight began the two drainages converged. The falcon was flying exactly parallel to the ducks despite being a great distance apart and not having seen the flock since leaving the snag. She had timed her route perfectly, and reached the confluence at the same time as the teal family. The falcon took the low position, forcing the ducks up. Faster than the teal, she began closing the gap between her and the birds on the northerly edge of the flock. The flock veered to the right, then panicked and turned left. The second to last duck at the north edge of the flock was late in reaction and barely made the adjustment necessary to remain in the formation. The bird in the outermost position took her cues from the second to last duck and wasn't able to make a similar recovery. She was now in the falcon's track. At the last second, the panic-stricken yearling chose to veer away from the flock, a fatal mistake. Having succeeded in splitting her off from the rest of her family, the falcon now made sure to keep her over the grassy mudflats. If the duck could reach water the chase would be over. With the falcon on her tail, the yearling tried a desperate maneuver and went into a nosedive. Seconds later the falcon slammed into the duck, grabbed one of her wings with her talons, and spun in circles. The centrifugal force made wreckage of the wing and the falcon let go. As the teal tumbled towards the ground, the falcon flew in a wide circle, meeting the duck in mid-air and grabbing her by the throat and abdomen, piercing her esophagus and lungs. The dying duck was a heavy load and the falcon's wing beats became fast and shallow as she lumbered her way to another snag, where she landed and began consuming her meal. Raptor took note of the falcon's capabilities and from this day forward would not venture over the mudflats.

Raptor's brother rarely encountered falcons, having moved inland and away from peregrine territory, though not because of them. He had found a long and winding riparian corridor, much of which ran through a regenerating forest that had burned six years prior. All of the old trees remained standing and healthy, their thick bark able to withstand the heat of the fire, but the young trees weren't so equipped and many had been killed. Six years later, wildflowers were exploding on the forest floor, as were berry-giving bushes of many types. The termite population had exploded, the queens having taken advantage of the fire's destruction. Almost every charred log, whether downed or standing, was home to a termite colony. The explosion of berries and insects in the forest brought those who feed on them, including small birds in high numbers. Raptor's brother had come upon a territory quite able to support a hawk of his kind. He established hunting routes throughout the area, learned the behavior of his prey, and perfected his hunting techniques accordingly.

Raptor, too, developed several hunting routes throughout her range, the day's conditions dictating which route she would take and when. Every so often she would awaken to the sound of migrating birds unaware of her roost site and could begin a pursuit right off. More often than not, however, Raptor had to spend the bulk of any day's effort trying to acquire food. Most days the hunt began the same, with Raptor dropping out of her roost and flying to the top of the tree where she would spend a moment listening for prey. If none were heard, she would head up Cattail Creek to a flat area above a waterfall called The Bench, where she would land on another perch and listen again. If nothing captured her attention she would continue up the creek a few hundred feet to Cedar Ridge, which was home to several of Raptor's prime surveillance posts. On a typical day she would fly fast and hard across the ridge, staying only a few feet above the ground until she reached the trunk of her fist post – a giant cedar over one thousand years old – at which point she would fly straight up and land on her favored branch. From this spot she could see the top of a vast tract of alders, the spruces and cedars of Wapato Swamp and Sitka Grove, a two-year old burn of several hundred acres on a hillside to the south, and the rim of the estuary for as far as her eyes could see. If no prey were in sight or earshot, Raptor would continue across the ridge, in the middle of which was a small clearing with salal and salmonberry bushes, and quite often sparrows and thrushes feeding on the berries. If the bushes were empty, she would cruise past and head to a three hundred foot drop-off at the end of the ridge. To reach the edge of the drop-off – and it's spectacular view of the valley below - Raptor landed on the ground, ran two-dozen feet down a raccoon trail beneath salal bushes and stopped at the cliff edge. By approaching the post on foot she would have a better chance of going unnoticed than if arriving by air. Raptor often spent upwards of an hour standing on the cliff and watching the happenings below. What she did next depended upon what she saw and heard. The majority of the time Raptor spotted small-bird activity in the valley and chose a stealthy approach through the under-story. She had close to a dozen routes into the valley and often cruised through them at blurring speeds, careening around tree trunks and branches whose location she had memorized. If she arrived at the cliff in the afternoon and found the valley empty of prey, she often flew down-slope and made a rather long journey to the roost by circumnavigating the entire estuary shore. The trip would of course be cut short if she succeeded in acquiring food. Sometimes she made an even longer trip of it and would work the north shore of the river before crossing it and coming back along the south shore, then continuing around the shoreline of the bay. If she arrived at the cliff early in the day and saw no prey below, she often flew upslope to a saddle with a pond in the middle. The saddle was a quarter mile long, wet year round, and home to many perching birds and dozens of red squirrels. There were a variety of these micro-ecologies scattered throughout Raptor's routes and the birds who lived there learned to keep an eye out for her, but because of her outstanding hunting abilities, the watchful eyes of her prey didn't hinder her food acquisition to any significant degree.

Raptor had a fruitful summer working her routes. She was well-fed, sparred occasionally with resident crows and ravens without serious incident, and was able to avoid the red-tailed hawk, who's hunting style differed enough from her own that they didn't come into contact often. Raptor grew quite large and strong for a hawk of her species and was well prepared for the cold and rain of the coming months.

Chapter 4

Autumn arrived as the bull elk were beginning their rut. The clearing beyond Sitka Grove had been the favorite battleground for untold generations of male elk and Raptor was awakened with regularity by their bugling, which was often followed by a sound like sticks hitting as the males engaged one another in antler wrestling. The matches were not intended to inflict damage on one another and injuries were extremely rare. However, the matches were of great importance as the winner would mate with many females in the herd, giving the species a survival advantage by ensuring the most fit individual contributed significantly to the gene pool of future offspring.

In the waters of the estuary, Coho salmon were jumping by the thousands. Some leapt out of the water to evade seals and sea lions, others to loosen the eggs within. Many jumped in preparation for leaping waterfalls, and still others jumped just to see the world above. On one occasion a small pod of Orcas came into the estuary and fed on the staging salmon, the evasive jumps of the fish marking the paths of the whales below. All six resident eagles and this year's offspring were also busy pursuing salmon. Raptor watched them fish now and then, occasionally witnessing a salmon leaping just as the eagle's talons hit the water, a very successful evasive maneuver for the fish.

A few weeks after the salmon arrived in their staging grounds the rains began. The prevailing winds of the Northwestern rain forests blow from west to east, often bringing with them moisture-filled air from the South Pacific. When cold air from Alaska and northern Canada pushes south and collides with the moisture-laden fronts, snow and hail happens, but this is a rare occurrence. This early-October day saw the more common weather pattern playing out, and a drizzle had been blanketing the estuary all day. While the rain on the immediate seaboard was light, it was a different story on the windward side of the foothills of both the Coast Range and Cascade Range. As the moisture-filled air gets pushed to the higher elevations of the mountains, it condenses, and the condensation drops as precipitation. So much water gets squeezed out on the windward side that the rain forests grow thick, and once the air reaches the lee side, there is little moisture left to fall. This occurrence is most pronounced in the Cascades, and the desert ecology begins with the descent of the foothills to the east.

As the heat and drought of summer wanes and cold and moisture once again claim the land, myriad mosses and lichens fill with water, doubling and tripling in size. On the forest floor, countless mushrooms of many types push up through the loam. The fruiting bodies of massive fungi living underground or inside logs, mushrooms drop spores, the fungal equivalent of seeds, and in so doing bring their essential contributions of decay to new areas. Many species of mushrooms are consumed by the mammals and birds of the woodlands, the king boletus being the most preferred. Growing up to a foot tall with caps a foot in diameter, deer and rodents often take bites out of the giant mushrooms leaving a perfect imprint of their teeth and jaw line. On one occasion Raptor caught a chipmunk while he was sitting atop one, nibbling.

Up to this time, Raptor had split her sleeping hours between three roosts, but neither offered satisfying protection from predators or the elements. Seeking a permanent roost, she was drawn towards the interior of the swamp. The interior offered protection from fierce winter winds, and though Raptor had yet to experience such storms, instinct told her to find a roost on the lee side of the swamp. Raptor had few predators to worry about during the daylight hours, and even fewer to worry about at night, none the less, finding a sleeping place safe from raccoons, great-horned owls and the tree-capable weasel called fisher was of concern. Beneath the twisted arm of a cedar Raptor found such a place. Years ago the limb of a neighboring spruce had come crashing down on the cedar, resulting in a two foot-long eight-inch wide twisting crack on the underside of a large branch a dozen feet from the trunk. The tree had responded by immediately sending sap to the area and strengthening the surrounding cell structure. In doing so the tree had healed itself, leaving the two foot-long crack to grow along with it. Raptor spotted the crack from below as the only entry was from beneath, and if she could access it, she would be largely out of range from any potential predators. While raccoon and fisher would be able to reach her, she would be concealed, and neither animal would venture forty-feet up and twelve feet out on a branch in their routine patrols. That her roost opened downward satisfied any concerns that an owl would reach her. Should her calculations be wrong and a predator attack the roost, the crack was wide enough that it would be possible to slip past, so up she went. After a few awkward flaps of her wings while hanging upside down, Raptor managed to get inside. She shuffled her feet, puffed her down and began to relax. A perfect fit. By day the young striker would travel her routes and perfect her hunting skills, by night she would find good sleep in the underside of a broken cedar limb.

Chapter 5

By the time of the winter solstice all migratory summer residents of the estuary had made it to their wintering grounds. Likewise, all winter residents of the estuary had arrived from their summer homes. For the southbound migrants, the central valley of California was a popular destination, but many flew even further south to the tropical forests and warm waters of Mexico, Central America, and South America.

Despite year-round food availability in the temperate forests and waters of the Northwest coast, the amount of food recedes along with the warmth and sunlight. However, because neither ground nor water freezes on the immediate seaboard, large numbers of ground feeding birds such as towhees, juncos and thrushes can be found scratching for seeds and insects year-round, with waterfowl foraging in the ponds, rivers, bays and ocean. For thousands of years the creatures of the estuary have lived in balance with this seasonal shift in food availability. Helping make food distribution more equitable, some species with similar diets leave as others arrive. Fish-eating loons arrive after fish-eating terns leave. Berry and insect eating Alaskan robins and golden-crowned sparrows arrive after flute thrushes and white-crowned sparrows depart. Many other species are part of a similar exchange. Not all migrants are replaced however, and species that catch insects on the wing – martins, swallows, nighthawks and bats – won't be seen again until spring. Nor will turkey vultures, the spring die-off and resultant carrion deposits having come to an end.

    Many birds of prey migrated through the estuary throughout the winter and Raptor generally tolerated them. Once she chased away a male striker who had stayed near Wapato Swamp for three days, one day longer than she cared to have him around. A gyrfalcon spent a few days hunting on the marsh flats and seashore, as did a rough-legged hawk. Raptor didn't harass either. Neither did she harass northern harriers, osprey or black-shouldered kites, many of whom lingered about the estuary for several days. The resident red-tail didn't share Raptor's tolerance, however, and he regularly harassed any bird of prey within close proximity to his hunting grounds, regardless of species or size.

One December morning Raptor poked her head out of her roost and for the first time in her life saw a snow-covered landscape. She spent a moment looking at the whiteness, and though she didn't know what it was made of, she simply accepted it and her curiosity quickly waned. The agile hunter stretched her wings and legs, then flew the short distance from her roost to the treetop. She spent the next several minutes sitting still, listening and watching, the route she would pick depending on what she saw and heard. On this particular morning there was no potential prey within eyesight or earshot, so Raptor flew in a direct line, fast and low to the tree line along the estuary shore, where she stopped to watch and listen again.

Far down the shore she saw five yearling crows picking through the wrack line. Raptor flew in their direction, staying concealed by the trees, then stopped nearby and watched again. Her opportunity for attack arose when one of the crows disappeared behind an old-growth drift log that had crashed into the river before being deposited on the shoreline during a massive winter storm five years prior. Raptor flew fast and hard - the other four crows screaming warning calls and taking to the air - and reached the log in seconds. She caught the crow on the ground and immediately began squeezing his neck and upper body with her talons. Raptor squeezed so hard one of his eyes ruptured. The other crows were in an uproar, dive-bombing and scolding her. It took fifteen minutes for the crow to become acutely starved of oxygen and his desperate heart to begin pounding erratically in his chest. As his raspy distress call began to weaken, Raptor started a rapid clucking sound that was commanding, assertive and hypnotic...*I have you! You relax and sleep! It is a good day to die! Be proud!* In a final burst of energy the crow began violently struggling and flapping, Raptor responded by calmly pinning the bird's wings to his sides and continued lulling him to sleep. *Cluk! Cluk! Cluk! Cluk! Cluk!*

While Raptor was eating the crow on the estuary shore, three thousand miles to the southwest an elderly amakihi was perusing a breadfruit tree for insects on an islet in north Polynesia. She saw the dark gray clouds on the horizon, white caps on the waves, and as the first leaf began to quake, the old bird knew ferocious times were on the way, and so filled her crop and moved inland. Two hours later the winds had reached sustained speeds of forty miles per hour with gusts approaching sixty miles per hour, and half of the tree's fruit had been violently thrown to the ground. In the open ocean a young albatross was gliding about, taking a calorie-free ride through the swirling tropical air.

    The first winds hit the estuary two days later, three hours after sun-up. To the east it was all blue skies and sun, but to the southwest the dark clouds from Polynesia filled the horizon. The sun was illuminating whitecaps and they shined bright against the dark gray sky, as did the spout of a humpback whale taking a breath, halfway finished with her journey from the Gulf of Alaska to Mexico's Magdalena Bay. By midday the blue sky was gone and a massive rain cloud hundreds of miles in length covered the land. The winds had reached sustained speeds of sixty miles per hour with gusts upwards of eighty miles per hour, and the rain not only blew horizontally, when encountering a rise, it blew up. In the open ocean, swells had already reached forty feet in height, and the storm had only just arrived.

    Drenching rains immediately raised the water level of the rivers and in turn the estuary, but it wasn't until the coastal mountains began to drain their loads that the flooding began. As the river quadrupled in girth and tripled in depth, giant cedars and spruces collapsed into the rising water from the eroding edges. Many of the giants washed all the way to the sea where they were tossed about like toothpicks in the massive swells. Some trees remained at sea for many years, occasionally scarring whales or offering shelter for small fish before being deposited well beyond the median high tide line during another winter's storm. Many washed-out trees didn't make it to the ocean at all, but ran aground in the estuary offering hiding and resting places for countless animals for years to come, including young salmon whose cold-water temperature requirements drew them to the fallen trees' shade. Small mammals living in the floodplain drowned by the thousands, becoming food for crabs, sturgeon and other bottom-dwelling scavengers of the bay and open ocean. Thousands of mice, moles, voles and shrews who were able to flee their flooding burrows became meals for savvy red-tailed hawks, marsh hawks, black-shouldered kites, great-blue herons and common egrets, who positioned themselves at the closest dry ground in order to take advantage of the chaotic exodus. While low-lying areas that constitute the floodplain would have a tremendous loss of rodent life, populations would rebound within months.

 Raptor was still expending energy she acquired after consuming yesterday's crow, but in order to remain in top condition, she needed to eat at least once a day. Winds of this extreme were new to Raptor, and she had trouble working her usual routes. The small openings in the foliage that she usually skirted at high speeds had given way to thrashing branches quite capable of killing her, so she flew above the treetops. With winds now sustained at seventy-five miles per hour, the treetop strategy didn't suit her for long. Raptor aborted the hunt for the moment and sat thirty feet up on a spruce along the south shore of the river. The sheer power of the storm captured her attention, and she could feel the land quiver as it was pounded by rain, wind, and waves.

Five hundred feet above the water and still a half-mile out at sea, the young albatross from Polynesia saw the headlands of the Oregon Coast quickly approaching and began desperately trying to escape the powerful winds and start a descent. In the millions of years of the species' existence, she was one of a small number to have ever seen the Oregon Coast, and the sight of the unfamiliar land made her uncomfortable. The albatross sought rest offshore, but couldn't get out of the winds in time and crash-landed in the flooded Wapato Swamp. The only albatross to have ever visited the swamp, she paddled about, regained her composure, and used the opportunity to pick at edibles floating on the swamp's surface.

The swirling storm was packing energy that, like a spinning bowling ball encountering pins, was being expended on the trees of a two hundred mile swath of the Northwest Coast, and they fell by the tens of thousands. Raptor's valley was absorbing the brunt of the storm's force, and as the strongest gust of the day – one-hundred twenty-nine miles per hour – encountered the tree in which Raptor was perched, it lost it's subterranean grip and crashed into the river. As Raptor took to the air she didn't dare open her wings to their full extent lest the bones snap. In a tucked position, she let the wind carry her a half mile away before she cached herself away in the middle of a huckleberry bush, well out of the way of the strongest gusts.

    A half-mile upriver anxiety was visiting a mother black-tailed deer and her two calves. Three days prior they had swum to a large island thick with spruce, shrubs and grasses. Having been born on the island they had easily crossed the river many times going to and from favored grazing sites. Today was of course different and the calves watched their anxious mother looking at the rushing water and could sense her insecurity. The rising water threatened to overtake the island in its entirety and much of it had already begun to resemble a swamp. As the water reached the doe's knees – and the bellies of her offspring - she decided to swim back to the mainland with calves in tow. The rushing water pushed them past their usual landing, which was blocked by flotsam anyhow, and with legs paddling fast and hard they aimed for the next nearest low-point on the opposing shore. When the doe was just a few feet from the river's edge a floating hemlock five feet in circumference overtook her, pushed her underwater, and she became trapped between two submerged limbs. When she couldn't hold her breath any longer she inhaled and her lungs filled with water. The two calves made it to shore unharmed and waited out the storm under the lowest branches of a young spruce, eagerly looking about for their mother who would never show.

The calves remained together for two years, foraging in nearby riparian areas, which included regular visits to the island of their birth. Upon their separation both deer found mates and reared many young for years to come. The body of their mother had washed ashore with the old hemlock, becoming food for a pack of coyotes, a fox, a mink, and countless flies and beetles.

Morning greeted Wapato Swamp with a calm indifference. A shrew came home late and slipped into his burrow unseen, having traveled the human equivalent of ninety miles while foraging for food in the ecosystems of five old-growth trees since sundown. As the shrew was fluffing his nest, the albatross swam across Wapato Swamp to its outlet at the estuary. She raised her wings, began flapping then running, and after thirty feet took to the air. She used the prevailing winds to her advantage, angled her wings and up she went. In less than a minute she was high above the estuary taking one last look at the mountains of the giant island that seemed to go on forever, then headed south, going back to the South Pacific and her Polynesian home.

A few hundred yards towards the mouth of the river two ravens and two eagles were sparring with one another for possession of a drowned raccoon. From a standstill, the ravens had the advantage, and harassed the more cumbersome eagles with enough tenacity that they eventually took to the air. The ravens began hurriedly feasting as the eagles soared upwards. After the eagles reached an altitude of a few hundred feet they concealed themselves from the ravens' view by hiding their massive bodies behind the treetop foliage. As the eagle's disappeared from view, the ravens became uneasy. Now the eagles had the advantage. Once the ravens' concern crossed a certain threshold, they left. Moments later one of the eagles went into a dive, came careening around a giant maple, and scooped up the drowned raccoon. Had the ravens remained feeding, it may have been one of them in the eagle's talons. The eagle flew to the mudflats with the raccoon, and was soon joined by his mate.

In the surrounding forests, a third of the trees one hundred years old or younger were snapped in half or felled completely. Many of the older trees lost their tops, but their bulk remained standing and they would live out their lives for another hundreds of years. Other small trees, mostly alders and hemlocks, remained rooted and had become intertwined with one another, as if braided by a giant. Logs of every size covered the marsh flats as well as the beaches above the median high-tide line. All of the dead logs, downed or standing, would become food or dwelling places for countless insects and animals for years to come.

Having spent the night safely in the heart of the huckleberry bush, Raptor returned to Wapato Swamp to find her roost gone and many other trees having fallen victim to the tremendous winds. Hunger pangs in Raptor's belly sent her out to work her routes, which she flew with caution, learning the changes the storm had brought to her territory. Within the hour she caught a wrentit and two hours later a purple sandpiper fell victim to her. With her stomach full Raptor flew to the interior of a spruce thicket to digest and rest. Thanks to instinct, stamina, and her outstanding ability to adapt to new stimuli, Raptor had survived her first winter storm.

Chapter 6

Throughout the estuary many markers of spring had begun to reveal themselves. The first plant to flower was the trillium, its three white pedals, spade-shaped leaves and short yellow stamen popping up in colonies throughout the forest floor. Skunk cabbage did the same in the wetlands, sending up a foot-long yellow stamen between giant green waxy leaves. Salmonberries and red-elderberries flowered soon thereafter, giving nectar to hummingbirds and bees, the resultant pollination giving life to the hearty berries within. Bleeding hearts were the next to bloom, their pink, heart-shaped flowers dangling from lacy green leaves also being a favorite of nectar eaters. High above the forest floor alder and maple trees were also budding, providing additional food for any who could reach them, being particular favorites of grosbeaks and band-tailed pigeons.

Innumerable insects emerged with the warmth, filling the air with life. As the insects took to the air, so came the insect hunters. Swallows arrived from southern latitudes – violet green, tree, cliff and barn – bringing with them an impressive array of aerial acrobatics and sweet-sounding chirps. In a matter of weeks, all cavities in trees and cliffs suitable for a swallow nest would be occupied. The nasally *beeep* call of nighthawks could be heard as flocks moved through, the giant brown swallow-like birds being particularly drawn to major hatches of termites and caddis flies. Misnamed on two counts, nighthawks are active in the day and night and are not hawks. Bats came too, from southern latitudes as well as nearby caves and tree cavities in which they had spent the winter.

Raptor's second spring was her first as an adult hunter. The vernal sounds, temperature variations, barometric fluctuations and precipitation patterns were becoming familiar to her, and she was learning to put these spring happenings to practical use. She learned that the calls of red-legged frogs preceded the arrival of massive flocks of northbound black-headed and evening grosbeaks by about three weeks. Huge numbers of the giant finches flooded tracts of alder and big-leaf maple, their single-pitch chirps filling the forest for miles. The largest of North America's finches, grosbeaks often hang upside down to feed, a trademark behavior of their parrot cousins who often share their wintering grounds. A few weeks after white-crowned sparrows began singing, turkey vultures showed up, Raptor noting their presence merely as a marker of time. About the time turkey vultures arrived, great flocks of Alaska and red-breasted robins departed. With the exit of the majority of robins to higher altitudes and more northerly forests came the arrival of other thrushes from the south, both species of which were taken as food. Some thrushes and red-breasted robins stayed in the estuary year-round but most were migrants, returning each spring to establish territories as close to one another as each would allow. Raptor would take full advantage of the presence of the thrushes as food, and their descending flute-like calls marked the end of many a day, letting Raptor know it was near time to return to her roost.

On a day of exploration, Raptor continued past a usual turnaround and didn't stop until she reached the end of a coast pine forest that butted up against a miles-long stretch of sand dune. She could see well out into the ocean and was particularly captivated by a colony of murres and puffins nesting on a basalt boulder three hundred feet offshore. Far away in the dunes a marsh hawk was dive-bombing and scolding a coyote. Moments later an adult bald eagle holding a five-pound salmon came flying upwind along the shoreline followed by her fledgling whose loud cries paid testament to the hunger within. Once satisfied with her chick's effort, the eagle dropped the fish to the hard sand below and it landed with a thud. The fledgling descended to the salmon and spent the next few hours on the beach feeding. Raptor soon headed back through the coast pine forest towards home, catching a song sparrow on the way. Once finished with her meal, she continued her journey around the estuary back to Wapato Swamp.

    As Raptor flew home, unbeknownst to her, a male striker saw her, and began to follow her from a great distance. Once she reached the edge of Wapato Swamp, Raptor landed on the topmost branch of a willow to rest. A minute later, and to her surprise, she looked up to see the male striker circling above. From a distance of four hundred feet, they made eye contact. The male – we will call him Darter - decided at that moment to pursue Raptor as a mate. There may have been a welcoming look in Raptor's eye, or perhaps he somehow knew he had arrived at the perfectly convenient time. Whatever the impetus, Darter felt welcomed enough and tucked his wings and dropped out of the sky, heading right for Raptor. At the last moment, Darter pulled out of the dive, and with his wings in a deep arc, began to fly in a circle over Raptor's head. As he made his way back around the circle and was once again in her vicinity, he fanned his tail, flashing her the black and white feathers on its underside. He then landed about forty feet from her and began to call out *cak! cak! cak! cak! cak!* His voice, like all male strikers, was of higher pitch than that of the females of the species, one of many indicators that female strikers are the dominant sex. Only after Raptor answered with a reassuring call of her own did the submissive male approach. He flew to within twenty feet and called out again.

Perhaps there was something unique in his display that pleased her greatly, or perhaps she would have welcomed any male of her species. Whatever the reason, Raptor again answered reassuringly and took off from her perch, beginning one of several slow-speed chases that would continue throughout the day. As the male pursued Raptor, he raised his wings high above his back and again flew in a wide arc around her, with slow, rhythmic flapping. During each pass he flashed his black and white tail coverts. After some time of this, Raptor descended to a favored perch and the male landed a few feet away. He began bowing towards his new mate, moving closer to her between each set of bows.

Later that afternoon, a feeding ritual began. Darter caught a wren, stripped him of his feathers, and laid the fleshy carcass on a branch in Raptor's sight and flew a good distance away to watch. Raptor made her way to the branch and began consuming the wren. Next, Darter caught a flycatcher and laid her on the branch without stripping the feathers. Raptor flew to the flycatcher but would not eat her and flew back to her perch. Darter returned to the corpse, removed the feathers, and flew back to his post. Pleased with the now featherless carcass, Raptor returned and began to eat.

The next day Raptor and Darter started building the nest. It began as Darter "crashed" into an alder, snapping a pencil-sized twig in mid-air. He flew with the twig thirty feet up to a crook in a magnificent burial tree and began to call out *cak! cak! cak! cak! cak!* These ancient burial trees stood throughout the rim of the estuary and held sacred import to the human inhabitants of the area. Many of the trees - including the one Raptor and Darter chose for a nest site - were more than five hundred years old as were some of the fallen relics that now lay decaying in the substrate. Upon the death of a tribal member, the survivors held ceremony and often "buried" the deceased on the tree. The body and earthly possessions were placed in a canoe and the canoe lashed crosswise, perpendicular to the branch. The branch was bent skyward at a right angle and tied to the trunk, rope holding the canoe in place. The lashing rope was made from carefully prepared stalks of stinging nettle and spruce root and would not decay until the branch altered it's cell structure and became permanently fixed in position. Most branches of a burial tree were used in this way and the curious right-angled branches circling the trees remained long after the bundles had decomposed.

All suitable branches on Raptor's tree had once held bundles, the last having fallen two hundred years ago. The slow gathering of a light mist eventually weighed just enough that the bundle came crashing out of the tree on an otherwise silent night. An ermine heard the bundle fall, as did a spotted owl and several of her rodent prey. After it had been on the ground for the passing of two moons, a bushy-tailed pack rat used one end for shelter and hay storage. His tenancy ended abruptly one night when a bobcat happened upon the fallen bundle and heard the rat within.

The bundle had contained the body and possessions of a fifteen-year-old hunter/fisherman named Standing Bear, who had been given the name after an incident involving a young black bear. His mother had left him on a bluff while filling seal bladders with water in the creek below, when a bear stopped beneath the bluff, stood on his hind legs and looked up at the two-year-old. Before the terrified mother could find her voice, and although the bear could have easily taken the boy, he slowly returned to all fours, huffed and ambled on. After the mother shared this story with the tribal elders, the boy was given the name Standing Bear. Before his death, Standing Bear had provided nearly half of the meat his extended family had consumed over the previous three years. He specialized in fishing by canoe, though during the peak salmon runs he would work the scaffolds set-up over the narrows on the big river and fish by dip net. He was an expert hunter of quadrupeds and took many elk and deer. Birds, too, were one of his specialties with turkey, tundra swan and several species of goose being his preferred avian prey. The loss of Standing Bear's food contribution impacted the family tremendously and the following winter was extremely tough, but they survived with help from the tribe. Standing Bear's many near-death encounters had become legendary and the stories were still told by elders around the fire. Over the past two hundred and fifty years the many tales about Standing Bear had morphed into cautionary lessons as he had died young and his death was such a devastating loss to the tribe.

The young hunter/fisherman had made his fishing line with kelp strands, spruce root and cedar bark, the leaders with doeskin or whale baleen, and the hooks with bone or stone. Lures were made of willow carved as small fish with pieces of shiny abalone shells affixed to give the appearance of reflecting scales. By wrapping the line around his hand while paddling the action of the lures mimicked that of live fish.

On the day of his death, Standing Bear had been fishing for coho and Chinook salmon at the mouth of the estuary where the fish were staging before their ascent to the mountain streams. The tribe held ceremony the previous day, celebrating the first deer and elk harvested by young tribal members. Smoked salmon, elk, deer, clams, sturgeon and smelt eggs had been served as the main course, with accoutrements of spruce leaf tea and berries on a clamshell. The celebration lasted all night long and there was much dancing, story telling and music. The next morning, as he was accustomed, Standing Bear awoke before anyone else in camp and headed by canoe to the mouth of the river to fish. The day's fishing was quite good and before midday he had caught two hundred pounds of salmon, the total weight of just eight fish.

As Standing Bear was pulling in his line and readying to head back to camp, an old brown bear losing his site began stalking a small band of elk on the marsh flats. Like the young hunter/fisherman, the old bear had spent the summer and autumn gathering food for the long winter ahead. However, he had failed to meet his caloric objective and would die during his winter sleep if he didn't increase his bulk. The bear had spent the day trying to catch salmon moving through a stretch of rapids he had fished throughout his life, but because of his failing eyesight, he didn't catch any at all, instead consuming the rotting carcass of a spawned-out fish on the bank. He had entered the estuary out of desperation, hoping to subdue one of a hundred harbor seals that were hauled out on the shore. His charge at the seals failed, and all one hundred of them crashed into the water with much splashing and carrying on. As the bear accepted his defeat and moved on, one hundred seal heads remained above water and watched him go.

The old bear was working his way back around the shoreline toward the mainland when he caught the scent of elk. Though his eyes didn't work, his sense of smell had not been compromised over the years. Being able to smell prey all around – yet unable to see them - had only accentuated his ornery nature. Making things worse, the old bear had been in a perpetual state of hunger since emerging from his den, with each day of hunger propelling his nasty nature to unimaginable new heights. In fact, things had gotten so bad for the old bear that on this day there wasn't a nastier bear in all the Pacific Northwest.

Standing Bear saw the elk at the same moment the old bear smelled them. The bear chose to approach the herd through a grove of stunted

Sitka spruce that had managed to colonize a strip of high ground on the flats. Standing Bear picked an approach by water, landing his canoe downwind of the herd in a slough that, unfortunately, was upwind from the strip of spruce. Standing Bear got on all fours, atlatl in hand, and began stalking the herd. One hundred yards behind him, the old bear began stalking Standing Bear.

Having been caught by surprise on the flats Standing Bear didn't stand a chance. The old bear was five times his size and able to run thirty miles-per-hour and he quickly killed the young hunter/fisherman, as he did two boys the following morning who were sent to look for him. After eating the three humans, the old bear had enough fat stored to make it through the winter. When he emerged the next spring, he lived for one month before being killed and eaten by a younger, healthier male of his same size. Having consumed the old bear of similar weight only a few weeks after emerging from his den, the younger bear's foraging was off to a good start.

At the end of two weeks, Raptor and Darter finished building the nest. When completed, the pile of twigs two feet in diameter and a foot high, lined with bark, and with a cup-shaped depression in the middle, was perfect housing for the young to come. Raptor laid five cobalt-blue eggs, and over the next five weeks she sat on them. The down on her belly provided perfect insulation, incubating the eggs at the exact temperature needed to allow proper embryonic development. During this time Darter hunted for his mate. Though he no longer stripped the feathers off the carcasses for her, he provided for her sustenance nonetheless.

Raptor left the nest to hunt on only one occasion, after a disabled purple finch flew past, listing a bit, then crash-landed on the ground a few hundred yards away. As Raptor sat on the ground pulling off the finch's feathers, a stellar jay noticed her vacancy at the nest site and descended towards the eggs with the intention of stealing them. If successful, he would not only provide nutritious food to his young, but also remove several potential predators from his territory. However, the jay was not successful in the slightest and instead became Raptor's next meal.

When the hatchlings arrived they were covered in white feathers and weighed one ounce, with a length of three inches. Raptor stayed on the nest with them for another two weeks while Darter continued his role as hunter/provider. As the chicks grew in size and approached fledgling status, Raptor helped Darter with food acquisition. Four weeks after Raptor joined in the hunting, all five young had fledged. Feeding five fledglings and teaching them to provide for themselves kept Raptor and Darter occupied from sunup to sundown. Throughout the forests and estuary, innumerable parents of many species of animals were doing the same.

Chapter 7

The flicker had been expecting the queens and knew exactly where to go after seeing the first rise up from the forest floor. With temporary wings giving her heavy body just enough lift to clear the tops of the brush, the queen had only a few hours to make it far enough to find a dead log of her own, where she could begin laying the eggs that were already developing inside her abdomen and thus begin a new colony. The flicker had fed on termites emerging from this log several times over the past three summers. There were several flushes during the warmest months, and she knew the exact holes from which the termites would emerge. Soon, queens began emerging from the log en masse, and in minutes three-dozen were airborne. The woods had been silent all morning, no predators evident, and the flicker was quick to descend upon the log, a termite flush being a good meal for any insect eater. Whether they knew it or not, one of the termites' strategies for the success of their species was to overwhelm predators by emerging in large numbers. That way, though many would be consumed and fail to breed, enough would be overlooked by the hunters that the survivors could establish colonies of their own. The flicker would never be able to catch every queen, despite positioning herself at the exit. And there she stood, jabbing her long beak into the hole and wiping her sticky tongue around the edges. In seconds she had consumed a dozen of the queens. Going back for a second round, she put her bill deeper into the hole, her tongue extending further into the decaying log.

  The collision nearly knocked her unconscious, and the pain in her chest and back was unfamiliar and unbearable. When she looked over her shoulder and saw the angry face of Darter staring back, her adrenaline ran thicker. Without hesitation she stabbed at his eyes with her bill - lightening-quick - but came up inches shy. Darter's long legs countered the advantage of the flicker's long beak and she couldn't reach his eyes, her only shot at debilitating the hawk. Darter had buried his talons into her flesh, and while they weren't in deep enough to kill her, the pain was searing. The flicker had a hold of the hawk's abdomen with her talons as well and the two were presently at a stalemate on the ground beside the log. The flicker was shrieking, spitting, and jabbing with her bill but coming up empty. Darter remained silent, patient and determined, and if all things stayed the same, it would only be a matter of time before the flicker would tire enough that she would relax her hold on the hawk, thus allowing him to reposition his hold and literally squeeze the life out of her. However, that would never happen.

Neither flicker nor hawk saw the six-inch paw as it smashed them into the ground. The flicker died immediately, but the hawk required further attention, and so the young mountain lion slammed his massive paw down again, this time crippling the hawk for good. So came the end of Raptor's mate Darter, leaving her the only provider for her family. After spending a few minutes smelling the dead birds and purring, the mountain lion began tossing them into the air one at a time, batting them about with his paws in a playful manner. Eventually tiring of his play, the young lion settled down to eat his prize. He had eaten two flickers before, having caught them on the ground while they were feeding on ants and termites respectively, but he had only seen hawks in the sky or perched in trees well out of reach. It was a bird nonetheless, and he ate it. Neither flicker nor hawk offered much in the way of food, and he would not have expended much effort in hunting them, and indeed he hadn't. He had been sleeping in the underbrush when the flicker had descended to the termite log, and was awakened by her shrieks when Darter attacked. He had moved only twenty feet from his sleeping place before killing the birds.

Raptor would live her life without any close encounters with a mountain lion. She would see several of them from safe distances while high in the air, but not once did any of them pay her any notice. Thrice she watched a mountain lion take a deer, and once witnessed a battle between an old lion and a young black bear, which ended the bear's life. Masters of stealth and camouflage, many more mountain lions had been in her field of view but she never saw them.

While Darter was stalking the flicker, Raptor had flown with two of her fledglings to a dirt cliff over a bend in the river. As she approached, a kingfisher began chattering warning calls before heading upriver to continue fishing on a safer stretch. Two-dozen holes in the dirt cliff were home to a cliff swallow colony, and it was toward them Raptor was heading. She had success extracting a chick from one of the holes a week earlier and decided to try her luck again. As the swallows bombarded her with voice and swoops, she grabbed onto the rim of one of the holes with one leg, and began reaching into the hole with the other. Inside the hole, the chicks silently pressed their bodies as closely to the floor as possible. They watched the long leg of the hawk with deadly talons sweeping around for them. To the chicks' good fortune, their parents had adequately excavated the nest hole and Raptor was unable to reach them. The harassing of the adult cliff swallows was an annoyance she didn't want to deal with, especially while engaging in a hunt with a low percentage of success, so Raptor headed to a nearby pond, followed by the two fledglings.

Raptor landed twenty feet above the water on her favored perch, and the fledglings watched in silence. In time, a red-legged frog emerged near the shore eyeing a damselfly that had landed on a low-hanging blade of sawgrass. As the frog crept towards striking range, Raptor crashed into the water, the splash silencing the singing of frogs on the opposite shore. The fledglings learned how to hunt frogs that day. After the fledglings ate the frog, the trio moved on, raiding a nest of thrushes, taking all the young and one of the distraught adults. When the strikers returned home they found the other three fledglings hungry and in a stressed condition. Raptor could sense something had gone wrong and knew that the three were hungry, but it had been a typically exhausting day for her and the sun was setting besides. The three would have to remain hungry throughout the night, unless Darter showed up with food before the twilight was extinguished.

Chapter 8

One of Raptor's three daughters was born with symmetric albinism on the tips of three tertiary feathers on both wings. The albinism provided neither advantage nor disadvantage, and her uniqueness was not recognized by any. For reasons having nothing to do with her white feathers she would become the most-fit hunter of the brood. At this point the family had accepted Darter's disappearance as final – it had been two weeks – and all five fledglings were dependent on Raptor for food and guidance. Raptor was having an extremely difficult time with her role as a single parent, but she succeeded in rearing them to adulthood, though it was to her own detriment. She had fractured her chest again, torn a ligament in her thigh, and received a hairline fracture on the radius of a wing. The radius fracture occurred when Raptor crashed into an elderberry tree, latched onto a band-tailed pigeon, and the two tumbled twenty-five feet to the ground. Upon impact with a rock on the ground, Raptor received the fracture. Despite the injury, with the help of pain-killing adrenaline pulsing through her body, she quickly subdued the pigeon, but that injury, like the others, would haunt her in the years to come. The band-tailed pigeon was the last meal four of Raptor's offspring would receive from her. They had learned how to hunt on their own and simply left over the next two days. Only the female with albinism remained, the unique white-tipped feathers having nothing to do with her staying. Raptor fed her only twice more, then no longer provided food for her, but she remained in the territory nonetheless, and often slept near her mother's roost.

    The hottest days of summer often ended with a chill as the fog rolled in. As the blaring heat of the high desert east of the Cascades quickly rose, it created a massive vacuum. The sucking action pulled all available air within five hundred miles towards the center of the hot zone. Though the burial tree was separated from the eastern desert by two mountain ranges and nearly two hundred miles, the vacuum pulled the offshore marine layer and it's thick fog inland, inundating Wapato

Swamp and the burial tree with wetness. The fog rolled down the valley at a snail's pace, moisture collecting on everything in its path. By the time the fog reached the burial tree, the surrounding woods were dripping wet. The mosses and lichens inflated with water and the salamanders and newts were roused from beneath the duff, and although it was imperceptible to the sentient creatures of the woods, the trees and other plants of the valley were thirstily taking the fallen drops of water into their bodies. While each droplet weighed less than a gram, the total water load taken into the plants of the valley then released back into the fog that afternoon weighed more than ten thousand tons.

Raptor left her nest to hunt in the thick fog. On this day it would be her ears that would alert her to the presence of prey. She flew to the interior of a spruce grove and listened. She could hear water trickling over the stones of Cattail Creek. She heard the croak of an approaching raven on a routine patrol, then the wings of two cutting through the air overhead. Far away she could hear the scolding calls of stellar jays, probably harassing the spotted owl who often hunted during daylight hours. The wind brought with it, from far, far away, three barks of a sea lion and a pair of crashing waves.

In time Raptor heard the approach of a large mixed flock of warblers, chickadees and kinglets. She flexed her wings and stretched her legs. Able

to decipher the flock's direction of travel, Raptor knew the route would dead-end where the forest came to a point at the edge of a high ridge and followed them. She generally stayed a hundred feet or so behind, but several times she flew very close to the flock and stopped just out of sight, letting them continue foraging unmolested, and when they moved off followed them again. Once the foraging flock reached the end of the woods, she would make her move. Cut off from the next tree line by a distance of a few hundred feet, she would attack when they flew across the large expanse. The chickadees were the first to go. One at a time, both chestnut-sided and black-capped crossed the distance. A myrtle warbler followed them, followed by two kinglets. The bulk of the flock was still in the tree, but it had been stripped of all insects in sight and wouldn't hold the attention of the remaining birds for long.
Suddenly, a flute thrush landed on the ground only thirty feet from Raptor. Having taken cues from the nonchalant behavior of the flock, the thrush didn't suspect the presence of a predator. About twice the size of a warbler, the thrush offered more food and was much easier to catch. The pursuit of warblers almost always required a chase through unfamiliar treetops at a great expenditure of energy. Thrushes were generally caught while on the ground and by surprise, no chase required. The capture of this thrush would be no different and

Raptor descended upon him, his shrieking hushing the forest and causing the flock of migrants to silently scatter. Though Raptor could have asphyxiated the thrush in a matter of minutes with her talons, she decided to enlist the help of a small pond a few feet away. She wing-hopped the short distance with thrush in tow and stood in the shallows, holding the struggling thrush underwater until he drowned. Once the flock saw the thrush secured in Raptor's grasp, they descended upon the shrubbery and low-hanging branches in her vicinity and barraged her with scolding calls. Raptor may have been annoyed by the raucous flock, but she didn't show it, and once she was finished eating simply flew off through the under story until she was out of sight, then perched to digest her meal.

Chapter 9

By early October much of the Northwest was in drought condition. The water volume of all drainages was at their lowest point of the year, but they wouldn't remain at those levels for long. The mosses had faded from vibrant greens to muted greenish-browns and had shrunk in size. The marsh grasses had matured and seeded, their bright green color now dulled. Most of the grasses' seeds fell into the tidal waters and were deposited in depressions in the mudflats and along the shoreline, creating feeding stations for ducks and geese until all available had been consumed. All breeding pairs of flocking species had joined the yearlings. By grouping, robins, blackbirds, chickadees and others have more eyes available for spotting predators and finding food. For Raptor, although her opportunity for stealth was reduced, the high numbers meant more prey were available per strike. This system of checks and balances had been established over millions of years to the advantage – and disadvantage - of both predator and prey.

One afternoon Raptor returned to Wapato Swamp to find a young black bear excavating a sleeping den beneath the burial tree. It was only a temporary abode for the bear - a place to sleep in close proximity to the great berry flushes about - and he would follow huckleberry flushes to the high ground and sleep away the winter months there. As the bear was digging, Raptor flew to a branch fifteen feet above him and watched. He heard her land on the branch, became curious, stood on his hind legs and looked up at her. The bear's curiosity quickly waned and he returned to digging. As the bear flung dirt from the depths across the leaf-layer, a stone was unearthed and it too was flung. The stone had a deep, smooth depression in the middle, worn over the years by humans who used it as sharpening tool.

While the black bear would take any meat he could find, he ate mostly berries this time of year. He consumed many pounds of salal berries and huckleberries in Raptor's territory and his dexterous lips were able to strip branches clean without doing any damage to the plant. He spent many hours each day foraging in the patches. One of Raptor's hunting techniques involved striking her prey while they were distracted. To achieve her ends using this method, she decided to shadow the bear during one of his treks. Raptor followed the bear to the slope above the bench where many birds also fed on the berries. When the bear ambled a bit too close to a towhee, the bird flew to a safe perch and began scolding him. Being preoccupied with the young bear the towhee didn't see Raptor coming and was quickly subdued. Having achieved immediate success, Raptor employed this strategy many times while the bear remained in her territory and caught many thrushes, sparrows, jays, towhees and waxwings while hunting in his shadow. Camouflaged in a forest of greens and browns, it was impossible for the average human eye to isolate Raptor's form at two hundred feet. For the average bird of the forest this was no problem at all, for the eyesight of any bird species is greater than that of any human. The bear, whose poor eyesight couldn't pick her out of the greenery at forty feet, almost never saw her. Once the cold winds and rains began, he moved upland to his winter den, and Raptor never encountered him again.

    The flowing winter was a tough one. Three unusual freezing spells kept the insects immobile and dug deep into cracks, crevices and burrows.

Up high the snow pack was abnormally deep and lasting, which meant birds that usually spent the winter there descended to the frost-free lowlands. The cold wasn't all bad news for Raptor, for while she had to increase her caloric intake, the arrival of prey from the highlands meant there was more food to be had down low. The wet and extreme cold, however, began to aggravate her injuries and arthritis took hold of her on many occasions. The usual lack of sun from cloud cover and inadequate solar incidence due to latitude meant cartilage and ligaments would lack sufficient amounts of vitamin D as a healing catalyst, and Raptor's wounds had trouble mending. But Raptor survived the harsh winter, her compromised condition giving rise to an inner strength of great dimension.

Chapter 10

Raptor bred each of the next six springs, mating with a different male each season. With the help of her mates, she successfully reared fifteen chicks to adulthood. Of the five young she lost over those years, one was taken by a Great Horned owl, one choked on a shard of bone, one fell out of the nest and died of hypothermia, and two were killed as fledglings by a goshawk. That more than sixty percent of Raptor's young made it to adulthood paid testament to her strength, tenacity and integrity. At seven years old, Raptor was in the upper middle age for her species. Due to her vast experience however, her mind was of elder status, and that, perhaps, was most responsible for her outstanding genetic contribution to her species. Her many injuries meant much of her body was aged more than was typical for a seven year old of her species. However, despite her many injuries, or perhaps because of them, Raptor was truly one of the most formidable strikers to have ever lived.

Raptor's last hunt began on a beatific spring day. Steam was lifting off the estuary waters, the wind was still, and a ray of sunlight gave warmth to her body as she stretched her wings and legs. After an unsuccessful hunt along the tree line on the south shore, Raptor was heading to The Bench when she heard the gathering of a large, mixed flock a few hundred yards up Cattail Creek. Migrating flocks of many kinds of birds often worked the creek, whose life-giving water attracted insects and allowed trees and their buds to flourish early in the season. The bulk of this particular flock consisted of migrating warblers, but many of the resident birds had joined into the boisterous chorus and feeding frenzy. Warblers of many kinds were flitting about from the tops of the trees to the lowest of the shrubs. Chickadees and kinglets joined them, and a flock of bushtits was also in the mix. Scattered throughout the undergrowth were two-dozen juncos and a half-dozen towhees scratching for insects and seeds. Raptor flew in a direct line six feet off the ground and slipped into the interior of a huckleberry patch without being seen. The birds were now all around her, flitting about and calling out boisterously. The close proximity of her prey caused a gland in her body to increase production of adrenaline. As Raptor waited for an opportunity to strike the adrenaline thickened in her blood and her heart began to beat faster. The adrenaline also caused her strength potential to increase as well as her overall awareness. Almost involuntarily she sprang from the branch.

    The jay had been hopping down the length of a spruce bough picking up bunches of lichen for his nest. A chickadee was the first to see Raptor and he called out, causing a chain reaction of both silence and scolding throughout the flock. The jay looked toward the origin of the scolding just as Raptor arrived. He flew in the nick of time and Raptor followed, inches behind. The jay flew directly at the trunk of a tree, then pushed off with his strong legs, and ricocheted in another direction. Raptor couldn't make the adjustment and slammed into the tree, but instantly regained her composure and kept up the pursuit. The jay was careening about the branches, going from tree to tree. By mixing flight with the ricochet maneuver, he was able to stay just out of Raptor's reach. A minute into the frantic chase, it came to an abrupt end. Raptor got a hold of him and they spun towards the forest floor. Raptor hit her wing on a branch and received the fifth fracture of her life. The pain was too much and she let go of the jay. The jay flew away, his flight a bit compromised, but if infection didn't get him, he would live. Raptor would not be so lucky and sat on the ground, stunned. The boisterous feeding frenzy had turned to silence, and the only sounds Raptor heard were the breeze in the treetops, the gentle ripple of Cattail Creek, and after several minutes of quiet, a lone tree frog emitting a single, raspy croak.

The pain in Raptor's wing was immense and she stayed near the ground for two days. She spent most of that time a few feet up in a salal bush, tried a berry once, spit it out, and started going hungry. By the third day the pain had subsided enough that she could fly short distances, but chasing birds was out of the question. On the fourth day, she half flew, half leapt to the scant remains of a beaver carcass left by a lone wolf on the estuary shore. This intake would buy her more time, but her latest fracture was compound, and bacteria began to replicate themselves en masse in her blood stream. As her body weakened, the torn ligament in her thigh began to irritate her, and the old chest fractures made themselves evident. The latest wing-break soon became extremely painful as the bacteria consumed her vitality, creating an infection that began destroying healthy tissue. The resultant fatigue lead to stress, which led to a compromised immune system, and soon a virus also took hold of her body.

Raptor clung to a low alder branch for three more days. As her vision became blurry and her head ached she began to experience auditory hallucinations. On Raptor's final day, her daughter with albinism landed next to her, pulled the feathers off a junco, and laid the fleshy carcass beside her. Raptor could see well enough to eat, and she did.

As her mother ate, the daughter flew off to the north, leaving the estuary for good. She eventually established a territory in a nearly impenetrable rainforest on the Olympic Peninsula, butted up against the ocean. She often worked the shoreline, which was curiously void of falcons, and often took sanderlings, sandpipers and plovers while they were resting among the driftwood. She watched as the smelt ran, the seagulls running about and plucking them from the shallows as seals, sea lions and porpoise dined on them beneath the breakers. From her roost she was awakened by the aurora borealis on three occasions. She dined on frogs often, shadowed brown bears on many occasions, and even spent several days hunting with a pack of wolves. She had found a truly magical spot and remained there for twelve years, successfully breeding each year but one. None of her offspring had any hint of albinism.

Raptor finished eating the junco, and later that night took her last breath. The following morning, her lifeless body was still clinging to the branch. A chickadee began scolding her, soon realized something was not right with the hawk and moved on. A moment later three native women and two girls entered Wapato Swamp, carrying baskets and telling jokes. They stopped at the edge of the water and began removing their shoes, then walked into the water up to their chest. Raptor had watched these women come into Wapato Swamp over the last several years, and the women had watched Raptor hunt throughout the estuary on many occasions. The capable hunter had even used them as a blind on two occasions, and five times took advantage of the distraction their presence caused in the forest to catch birds; three Alaska robins, one ruffed grouse and one black phoebe.

The women were presently after the tuber of the Wapato plant, an essential food for the tribe. The leaves and flowers of the Wapato remain above water while the root system remains buried beneath the pond and the women were using their feet to harvest the tubers from the mud and having a great time doing so. After filling their baskets with the tubers the women prepared to leave. One of the girls was the first to spot Raptor's body.

"Mother, is that the hawk?"

The mother walked beneath the branch and stood as sadness came into her eyes.

"Yes, that's her."

After uttering a short, almost silent prayer, her attention was drawn to an object on the ground and she bent over to pick it up.

"Look!"

In her hand was the sharpening stone the young bear had uncovered several years prior. As she wiped off dirt and a carpet of moss that had grown on the stone, she saw affixed to it an abalone shell inlay of a bear standing on his hind legs. She looked up at Raptor.

"It's like I said. This great hunter held the spirit of Standing Bear."

As the women left Wapato Swamp, a young striker entered the skies overhead. The hawk had been born in the foothills of the Sierra Madre mountains of west-central Mexico, had flown north to the arid deserts of Southern California, found the habitat disagreeable, flew west over the Sierra Nevada mountains, north over the Siskiyou mountains and was now circling over Wapato Swamp at one thousand feet. The lush forest, high ridges and cascading creeks reminded her of the forest of her birth. Hunting in this habitat would be familiar, an advantage she did not have in the Sonoran Desert. Then, with the angry look affixed to the faces of all birds of prey, the young striker began her descent towards Wapato Swamp to investigate.

The End

Andy Norris is a writer and filmmaker living in the Pacific Northwest.

Films:
Source To Sea: The Columbia River Swim
Targeting Iran
The Wolf Watchers

Books:
The Dancer Diaries
Raptor!

Photo Credits:
Mountain Lion, Duncan Parker
Cooper's Hawk Chicks, Tom Muir
Individual Hawks, Betty Norman

Made in the USA
Middletown, DE
10 June 2021